GREENERGIZED

GREENERGIZED

Dennis Posadas

Routledge
Taylor & Francis Group
LONDON AND NEW YORK

First published 2013 by Greenleaf Publishing Limited

Published 2017 by Routledge
2 Park Square, Milton Park, Abingdon, Oxon OX14 4RN
711 Third Avenue, New York, NY 10017, USA

Routledge is an imprint of the Taylor & Francis Group, an informa business

Cover by LaliAbril.com

British Library Cataloguing in Publication Data:
    A catalogue record for this book is available from the British Library.

ISBN-13: 978-1-906093-88-4 (pbk)

For Vicente and Mercedes

1

The students of Oriental College looked just like those in any other college campus. It was summer. In blue jeans, t-shirts and sandals, they were scattered around the courtyard reading, eating and gossiping. Midas Auditorium was at the end of a long hallway, past the courtyard on the right where the Venus de Milo statue stood in a small pond teeming with tropical fish. The sun, source of boundless energy, blazed over the rooftop, casually ignorant of its relevance to what was happening below.

"…so, given that anthropogenic climate change is irrefutable, solutions to reduce fossil fuel use are an immediate priority," intoned the speaker on the podium.

This was Professor Ruiz, a familiar figure on campus. With his blue long-sleeved shirt and khaki pants, coupled with brown horn-rimmed glasses and pepper-grey hair, he had stepped straight out of a '70s TV sitcom. Some of the students in the auditorium lolled around at the back, engaged in nothing but their own chatter. But there were a few attentive listeners up front, asking polite questions.

"But clean energy is something that will only be practical a few years from now. Yes? No! Some of those clean technologies have already reached grid parity," said the Professor. Sensing he'd lost at least two of his audience, he added, "Grid parity, for those of you who don't know, is when the cost of electricity from renewable or clean energy matches that of typical fossil fuel sources used by electric utilities."

But then there was a loud voice from the right side of the auditorium: "But how's that work, sir? Solar costs two dollars a watt and coal only costs a few cents.[1] The fuel companies have been cleaning up their act, haven't they? And besides which, they've been working with fossil fuels for decades, so their technology's more advanced – isn't it?"

Professor Ruiz screwed up his eyes to make out the owner of the voice. "That's actually a very intelligent point. What's your name?"

"Daniel, sir."

Daniel Vargas was an intelligent young man who had sailed through high school on a raft of top grades and high expectations. His parents were poor and had sacrificed a lot to get their only son the education that he was so clearly cut out for. He dressed rather conservatively for a young man of twenty-one: casual-smart short-sleeved shirt and chinos with leather shoes. His brown hair was neat and short, with a side parting. He had developed an arrogant self-confidence about his own intellect and was prepared to argue with anyone on most subjects.

"Well, Daniel …" began the Professor but Daniel hadn't finished. "How can people say renewables are competitive, when they obviously cost a lot more? Frankly, I don't get it," continued the student.

---

1. Approximately $2/watt in 2009 for photovoltaic solar; the figure for thin-film solar is lower, and was at the $1/watt mark during the inauguration of Barack Obama. This cost does not include land and installation costs. Obviously, technology and increased adoption will drive prices down even further.

"Well, Daniel, I actually agree with you. Some renewables *are* still expensive at the moment. But" – he stressed the word "But" with a wag of his index finger –"But! If government implemented the right incentives and subsidies, it *could* compete with fossil fuel-based power. Technology is advancing week by week to address cost and efficiency issues. And what is usually missing from the equation: you have to factor in the social costs of added health care, the impact of climate change and pollution of the environment."

Daniel set his face as he patiently listened to Professor Ruiz explaining that there were many types of solar cells. Some, like silicon-based photovoltaic cells, were a bit more expensive but were more efficient. Others, like thin-film photovoltaics made of materials such as cadmium telluride, were cheaper but less efficient.

Nodding in an unconvinced way, the student smirked, and then quite rudely muttered to his classmates: "The guy's a tree hugger: all that stuff about renewables being ready for the market is rubbish. I'm not letting him get away with this." Daniel drew himself up to his full height: "Sir, all of this is very nice to hear, but the fact is: these technologies are still really expensive. We can't afford them. In the meantime, the fossil fuel and coal industries are cleaning up their act. So isn't the sensible option to stick with what works and improve it rather than some pipe dreams of hippies and liberals?" In a slightly lower voice he added, "…and university lecturers?"

The effect was immediate. As if a whip had been cracked, students emerged from their conversations or slumbers and looked at the professor expectantly. Professor Ruiz cleared his throat and spoke slowly and deliberately: "There are ways that governments around the world have made clean energy attractive to investors. For example, there are, aside from the regular incentives like tax breaks and credits, new incentives like carbon credits and feed-in tariffs. There are also voluntary carbon markets, where companies and individuals can purchase carbon offset credits voluntarily even without a mandated cut. Increasingly, this has been born out of a trend where companies want to claim they have zero carbon emissions, so aside from their internal reductions, they can also buy carbon offsets from the voluntary market…" He trailed off, noticing that his efforts had done little to alter Daniel's skeptical expression.

At this, a female co-ed advanced towards the microphone. She was uncommonly pretty and momentarily distracted Daniel, and most of the other boys, when she spoke. "My name is Maria, sir. I'd like to ask you what you mean by a feed-in tariff?"

After a pause, Ruiz gratefully continued. "Yes, thank you, Maria. Good question. A feed-in tariff is an example of a new type of incentive to encourage investors to set up clean energy plants. Incentives like feed-in tariffs and renewable portfolio standards mandate that electric utilities purchase a certain percentage of their electricity from clean energy sources by a certain year. Now there are slight differences between these incentives. The feed-in tariff, for example, normally mandates electric utilities to purchase the renewable electricity at a higher cost as compared to fossil-fuel-based electricity. In turn, the utility can then pass on this added cost to consumers."

Feeling like he'd won the room again, he removed his horn-rimmed glasses and placed them in his breast pocket. "Now, since the utility has thousands of customers, this added cost can be spread out among all of them, so it only shows up as a minor addition to the electricity bill…"

Daniel leapt up excitedly. "That's what I mean! You see, it *is* still an added cost to the consumer!"

The Professor was ready for this one: "Technically, yes, of course, it is still more expensive at the start. But you have to remember that in order to spur massive adoption, which leads to economies of scale that will drop the cost of renewables in the future, sometimes governments have to enact these types of mandates. Fossil fuels, because they have been the de facto standard for many decades, have already reached the point of being mature technologies. Of course, there are still going to be new ways of cleaning up fossil fuels, but things simply cannot go on as usual without catastrophic damage to the environment. In fact, the fossil fuel industry itself receives large subsidies worldwide. So it isn't the free and fair market you may have been led to believe," he said, playing, as he saw it, his ace.

2

Professor Ruiz paced purposefully across the auditorium in what he liked to imagine was an authoritative manner. The atmosphere was different now, after this confrontation. Students who had been slumped lazily in their wooden seats straightened up. Some were taking notes; others were content to listen.

Still pacing, Ruiz continued: "In addition, there are other incentives that also rely on early adopters. The cap and trade system, for example." He scanned for expressions of comprehension, of which there were few. "In the cap and trade system, the government sets a maximum ceiling – a cap – for carbon emissions which a country cannot exceed. To enforce this, it sets up a trading system," he said patiently and deliberately, emphasizing *trading system*. "In a cap and trade system, those companies or entities that spend on carbon reduction projects get credits which they can then sell to other companies that did not make the necessary investments to enable them to meet the maximum carbon targets.

"And! Aside from the carbon markets, which operate because of mandated cuts, there are now *voluntary* carbon markets as well, where corporations and individuals can purchase carbon offset credits voluntarily – without anyone forcing them to do so."

He sauntered back to his podium and took a gulp from his glass of water. For the first time he noticed the heat and the damp under his armpits. He folded his shirtsleeves up to his forearms. The class was quiet, but still attentive.

"The carbon cycle!" he announced, rhetorically. "Unless the amount of carbon we generate falls to less than the amount that the oceans and the plants can remove from the atmosphere, we are going to see a spiraling rise in global temperatures and a corresponding rise in ocean levels." He explained that the Intergovernmental Panel on Climate Change (IPCC) had concluded that climate change was accelerating because of a lack of significant action in cutting back fossil fuel emissions.[2] "We are already seeing the melting of some polar icecaps and the thermal expansion of seawater, and this will cause sea levels to rise, threatening countries and places in low-lying coastal areas," Ruiz said. "This is not theory. Especially not for the inhabitants of Pacific Islands. The low-lying Pacific nation of Kiribati is negotiating to buy land in Fiji so it can relocate islanders under threat from rising sea levels. Some of Kiribati's coral atolls are already disappearing beneath the waves."

He paced across the room, and clasped his hands together while staring earnestly at his class. "In fact, even if *all carbon dioxide emissions* were to be halted today, the changes that are happening in surface temperature, rainfall and sea level may no longer be reversible, at least not for a very long time," he announced dramatically.[3] "Plus we are nearing 'tipping point' levels of greenhouse gases in the atmosphere."

---

2. See the website of the Intergovernmental Panel on Climate Change at www.ipcc.ch.

3. See the US National Oceanic and Atmospheric Administration (NOAA) website at www.noaanews.noaa.gov/stories2009/20090126_climate.html.

As the impact of his words fell, with fortuitous timing the bell rang. As the students filed out he casually announced: "Those of you who wish to attend the second part of my talk: tomorrow, 2 pm, same venue."

3

Deep in thought, José Garcia stared out into the city's business district through the window of his 35th-floor office. José was CEO of Acme Oil, one of the largest oil companies in the country. Today's concern was a certain Professor Emilio Ruiz. Ruiz was a prolific media commentator with a single agenda: the criticism and calling-to-account of the global fossil fuel industry. And a vociferous proponent of renewable energy. Furthermore, he was widely read, and widely quoted. The time had long passed when he could be dismissed as a crank. Up to now José and his advisers had seen this as a marketing war – positioning their point of view more strongly in the marketplace, where they must surely win. But the best PR agencies that money could buy were not having any effect in discrediting the Ruiz agenda. José was ready for new ideas.

José was one from the old school. One of the old guard in the oil industry who would have no truck with renewable energy, and saw it as nothing more than a threat to their existing business. Most oil companies were already widening their options and looking at renewables. Many had subsidiaries performing R&D in clean energy. But not Acme Oil.

"It's not working," he said, stabbing a finger at Professor Ruiz's guest editorial in the *National Journal*. "This guy is causing us real reputational damage! Why are they all giving him a platform for this nonsense?"

José's right-hand man and de facto PA was a slight middle-aged man with a pinched face. Unmarried and living alone, he was a company man who did his boss's bidding quietly and efficiently, often working long hours and weekends. "Sir, it's my opinion that we need a star performer of our own. Young. Articulate. Good-looking. Someone who is *actually* a passionate advocate for the oil industry. Someone who can go on the discussion programs and give Ruiz a run for his money."

José breathed out slowly as he considered. "OK, Miguel. Get the HR people on to it," he demanded.

"No need for that, sir. I've already found him."

4

Daniel was about to sit down to dinner with his family when their landline rang. "May I speak with a Mr. Daniel Vargas," said a cold official voice. "Speaking," he replied.

"Daniel, this is Miguel Contreras of the Acme Oil Company. I hope I'm not calling at an inconvenient time. I have a message from Mr. José Garcia, our CEO. He would like to invite you to lunch tomorrow. Valdez Restaurant, downtown on Bolivia Street, 12.30 pm. I should also tell you in advance that Mr. Garcia is looking for a successful candidate to fill a position in his company."

Daniel was speechless for a few seconds and wondered whether it was a prank call. Dismissing this possibility, he found himself saying "Wow!" Then, "Yes. Yes, of course. Of course. 12.30. Valdez Restaurant. Thank you. Thank you very much…"

"Mom. Dad. I've only just been invited to have lunch with the CEO of Acme Oil," said Daniel, more bemused than excited.

His father smiled. "You'd better borrow my suit."

5

Daniel made a point of being in the bar area of the restaurant fifteen minutes before the appointed time. He felt awkward in his father's suit, and wished he'd finally gone out and bought one for himself. He scanned the entrance nervously every couple of minutes, and at 12.29 he saw a distinguished-looking man in a grey suit get out of a limousine. The man, who looked younger than Daniel had expected, walked towards him smiling. "Daniel?" said José, extending his hand.

Daniel shook it. "Yes, sir. How do you do?"

"Call me José."

At the dining table, Daniel was willing himself to stay cool and find some *savoir faire*. He played it safe and ordered what the CEO had ordered.

"Daniel, you're wondering why I invited you for lunch. Allow me to cut to the chase." Daniel wasn't used to being told what he was thinking. José continued: "We are being skewered in the media by Professor Emilio Ruiz. You know him. I'm not saying all his arguments are wrong, but he has a large sympathetic readership and the whole debate is … is losing balance. It's all too black and white and we're losing public trust. Simply saying that fossil fuels are a dirty energy source … well, it's not even true anymore. The industry's changing. There's been a *lot* of R&D on cleaner technologies."

Daniel said nothing, just nodded.

"You've been noticed, Daniel. I have a role for you. I'd like to see you draft some written responses, verbal ones too, to Ruiz's arguments."

Daniel's mouth was slightly open and without realizing it he had stopped eating. His usually quick mind was laboring to process what was happening.

"A permanent position is available if I like what I see."

Daniel had a hundred questions but failed to find the words.

"If you can't make a decision now – fine. I can wait a few days."

Daniel regained his composure, and lowered his knife and fork. He dabbed his lips with the napkin and said, "I'm in. When do I start?"

6

Maria was staring at her laptop. She was reviewing the draft of the guest editorial that Professor Ruiz had asked her to type up and polish. He was beside her peering at the screen. "Er, Professor, don't you think we should be actually saying what the carbon credit mechanism is?"

"OK, let's put it like this. 'Carbon credit is a monetary payment that is made in exchange for reducing carbon emissions. The typical basis for payment is one ton of carbon dioxide emission reduced. Some gases or substances have stronger climate effects, and therefore the price per ton for those gases is a multiple of that of carbon dioxide.'"

"I don't think that really explains it completely," said Maria. "How can we explain the fact that the valuation of the carbon credit is based on the equivalent volume of CO_2 that the project could reduce in atmospheric emissions? Plus, it doesn't explain the fact that these credits are traded in a market, so the price fluctuates. And it doesn't mention that normally carbon credits only pay for a fraction of a project cost upfront."

"Well, yes, you're quite right, Maria. But you can't squeeze all that into one editorial. This has to be one of a series of articles. Educating the public is a gradual process. Just work on it some more tomorrow and see where it takes you.

"On the subject of which … we'll need a similar piece on cap and trade."

"I'll do that now if you like?" Maria offered helpfully.

"I thought you had a date tonight? Go on – it's getting late. I should be getting back home, too," said the professor. Then deciding he would be nosy: "Who's your date with anyway? It's not with that young Daniel guy, is it? A bright boy. You could do worse, but oh boy does he need to grow up a bit and broaden his outlook."

Maria was pulling a face. "Not in a million years."

7

José's wrinkled brow was buried in a print copy of the *National Journal*, specifically the Editorial, the headline of which read: THE END OF THE FOSSIL FUEL ERA.

Daniel had already read it online on his phone but waited patiently until José was done. He tossed the newspaper casually onto the table. "This is exactly what I mean, Daniel. We're allowing this dangerous nonsense to go unchallenged. I'm not saying forget renewable energy entirely, but it's only ever going to be marginal – 1%, say. At the most. Oil is what has kept this country running for decades, and oil will still be the one that keeps it running for many more decades to come. We know the reserves are still out there. Renewables are too expensive and will stay that way." At this he stared directly at Daniel. "That's your key point."

"I think I already know how this piece should go," said Daniel.

Miguel had been thinking, fingertips together, lips pursed. "Maybe …" He drew out the word *maybe* until both José and Daniel were attentive for the completion of the sentence. "Maybe we should be doing more than just guest editorials. *Maybe* we should go for a televised debate," he proffered.

"Will the TV stations go for it?" asked José.

"Energy is a hot topic right now. We've just come out of a global oil crisis," said Miguel. "It's just a case of talking to the right people, pulling the right strings. Making them think it was their idea in the first place."

José nodded. He motioned them to help themselves to the snacks. Daniel asked, "Do you see someone going head-to-head with Professor Ruiz? Who's going to represent us?"

"You are, Daniel," said José calmly.

"Me, sir? Why me?"

"This is exactly why I hired you. You're young; you're clever. You'll appeal to a whole new demographic. And we know you can give Ruiz a run for his money."

Miguel had immediately started to make some calls. After a few minutes, he nodded. "It's done. It'll happen sometime next month. But we still need to get the invitation out to Professor Ruiz once the station gives us the mechanics," said Miguel. "And get him to accept."

José asked about the cost. Miguel explained that the TV station would not want Acme Oil to sponsor the debate in order to maintain a sense of impartiality.

José smiled and gazed out over the city. "I think this is going to be something to look forward to," he said.

8

"You are not going to believe this, Maria," said Professor Ruiz, a letter in one hand, a torn envelope in the other. TV3 want me to go on a live debate with Acme Oil."

"What's wrong with that, Professor? You've done debates with other oil industry execs, haven't you?" quizzed Maria.

Professor Ruiz silently handed her the letter.

"What? Daniel Vargas? How … ?" she trailed off.

"It looks like I underestimated his allegiance to fossil fuels. What is Acme Oil *doing*? He's just a kid. Well, I can't turn it down – I'll have to do it." While he was instructing his secretary to call the TV station to confirm his appearance, the phone rang. After a brief exchange, the Professor turned and beamed.

"Well, well. What do you know? That was Congressman Suarez. He's going to sponsor a bill on clean energy. With the Feed-in Tariff and the Green Option. It's as if he's using my article as a template," enthused Professor Ruiz.

"That's great news, Professor!" said Maria. "But even if it becomes law, won't there still have to be some type of massive information campaign to tell people what these things are?"

"Oh, yes, absolutely," said Professor Ruiz. "Case in point: I'll be talking at a Parent Teacher Association meeting this evening. They've asked me to talk about the new clean energy concepts. So in describing the Green Option, for example, I'll tell them that in the future people will have a choice of energy billing plans just like they do now with their cell phones.

"In a society where there's a mix of traditional and clean energy technologies, customers can go for the cheapest option, which probably means using traditional sources. But there is a green option: some might be willing to pay extra to receive electricity coming from clean sources.

"It's going to cost people a little extra to pay for green energy, but for some citizens it will be their way of investing in their children's future. And that's why it's such a good sell to the PTA."

"Wasn't there something about smart appliances and net metering?" Maria asked.

"Yes. Basically, what it is: smart appliances will have microchips in them to give them intelligence. So, for example, if a utility customer opts for the green plan, then he or she can set the refrigerator or air conditioner to go on a power-saving mode during the times of the day when there is peak demand, and go on the full power mode during times when there is less demand.

"And utilities will also be able to control these appliances in the future: for example, turning off coffeemakers that have been running empty for hours, or lowering the thermostats of air conditioners remotely. That is a way for the power companies to avoid building more polluting power plants, by making energy conservation technology-driven rather than just a voluntary activity.

"I can make an analogy with the food catering business here. One of the inefficiencies of the catering business is that caterers have to deal with the fact that their clients can't accurately estimate the number of people who will be coming to a party. Some of these people may be dieters, while some might have large appetites. So caterers typically allocate a lot of extra food, which typically gets wasted. In other words, it's a case of poor demand-side management.

"It's the same in the power industry. Electric companies have to build plants to meet the largest electrical demand that they expect. They also have what is called a spinning reserve, meaning that the generators have to keep spinning and emitting carbon even if they aren't supplying any demand at that moment. And a lot of this demand is not used, or is wasted energy. For example: air conditioners that keep the rooms very cold.

"If there was a way we could turn off, or control these appliances that are wasting energy, then we won't need as many power plants as we do now."

He pointed to an incandescent light bulb at the far end of the room. "That type of bulb, an incandescent lamp, wastes most of the electricity through heat instead of light. Older bulbs should be replaced with CFL bulbs[4] or, better still, newer types like LED[5] lighting."

"So what about net metering?" asked Maria.

"Yes, I think that should raise a few eyebrows: allowing your electric meter to run backward or forward. Soon you'll be able to sell power to the power company, instead of just being its customer. In California and Europe, companies – and individuals – are now beginning to install solar panels, biomass systems, micro-hydro systems, wind turbines to generate their own power. These people will have the excitement of receiving checks from the power company at the end of the month, instead of the money always flowing the other way."

Maria smiled confidently. "I don't think Daniel stands a chance," she said.

---

4. Compact fluorescent light bulbs, which save energy.
5. Light-emitting diode.

9

The TV company had pitched the debate as nothing less than a showdown between the fossil fuel sector and the clean energy sector. The hype had worked and there had been a flurry of media excitement.

The moderator of the debate was a well-known talk show host. She began: "Ladies and gentlemen, the guest on my right is Professor Emilio Ruiz, Gabriel Espinoza Chair in Applied Science, Oriental College. One of the most outspoken critics of the oil industry, he has been speaking and writing on behalf of clean energy causes for some time now. In fact, he has been a guest on my program several times. Welcome back, Professor Ruiz."

"Glad to be here," said Professor Ruiz, dressed in his familiar blue long-sleeved shirt and khaki pants.

"And on my left a more unfamiliar face. A recent graduate of Oriental College, in fact a former student of Professor Ruiz, a new recruit for Acme Oil and tipped for great things. Let me welcome Daniel Vargas to the show."

"Glad to be here," said Daniel, echoing the Professor. Today Daniel was wearing a designer suit and feeling all the more confident because of it.

The host explained that, instead of a formal head-to-head debate, it would be a talk-show-style discussion where she would raise a question and either of the guests could respond. The host might also ask specifically directed questions to either of the guests, and the other guest could react in turn. Having explained the rules of engagement, the host fired off her first question.

"Daniel, my first point to you if I may. Why are you in favor of fossil fuels being the basis for the bulk of most of our energy needs?" Daniel smiled. "I'm in favor of fossil fuels, and technologies like nuclear and coal, because, frankly, the technology is mature and widespread, and a lot of scientists and engineers are developing technologies to make it cleaner and cheaper."

He explained that fossil fuels were being attacked unjustly in the media, and that there had been a lot of progress in cleaning up emissions. He pointed to catalytic converters, to unleaded and cleaner-burning gasoline, to the use of biofuel additives, cleaner coal, and compressed natural gas – all examples of technologies developed by the fossil fuel sector to clean up its act.

"While there is a role for renewable energy, it is in its infancy and it is still too expensive and unreliable. In the meantime, we still need fossil fuels to generate energy for most of our electricity and transportation needs," he said, "without which society as we know it would simply collapse into chaos."

"And Professor Ruiz, how would you respond to that?" asked the host.

The Professor removed his glasses, folded them, and waved them as he spoke. "It *is* true that we need to rely on fossil fuels at the moment for most of our energy needs. It *is* true that it is cheaper, but it is cheaper because we have *let* it be cheaper. It has become a self-fulfilling prophecy, where we always say that renewable energy is more expensive, and therefore do not take steps to invest more in R&D, to encourage incentives to increase early adopters."

The Professor explained that the more people bought and invested in renewable systems, the more the price would drop because of manufacturing economies of scale, and a renewed salvo of investments. "That is not happening simply because the proponents of fossil fuel technologies have the loudest voices," he said.

He added that mandates in California and in Europe had caused renewables to approach grid parity, where the cost of the renewable system generated electricity approached that of traditional fossil fuel grid electricity.

"But, Professor, you must admit that governments have had to spend large amounts on subsidies for these renewable energy systems in order for them to become competitive. Without those subsidies, they wouldn't stand a chance," said Daniel.

"True, but isn't that government's role? To sometimes step in and create opportunities where new technologies and means of solving problems can take root?" queried the Professor rhetorically. "Fossil fuels might be cheap, but they have hidden costs in increased pollution of the environment and increased health care. Plus, fossil fuels are also heavily subsidized, so why the hypocrisy? And let's not forget the recent oil price crisis."

Daniel let the comment on fossil fuel subsidies go, but quickly pounced on the oil crisis comment. "These things happen in a market, Professor, but look: prices have stabilized back to their previous levels. Yes, sometimes oil prices go up – but they're low again now. Plus, renewables can be intermittent. The sun doesn't always shine, and the wind doesn't always blow."

"Our country has spent a lot on importing oil," countered Ruiz. "What did all that money get us? If we spent even a fraction of that cash on renewables, within a certain period we will end up with an energy source that does not need to pay for future costs – unlike fossil fuel sources. While the initial cost of a renewable energy system might be higher, there is almost no additional cost moving forward. Sunlight, geothermal, wind, even biomass energy are free or are things we take for granted or throw away even."

He faced the audience now as he explained that solar and wind systems were often connected to the grid, or to systems like batteries or other types of energy storage like elevated pumped storage lakes to compensate for their intermittent nature.[6]

Daniel was losing his cool. He knew that he'd never stay in favor at Acme Oil if he lost this debate on national television. He blurted out, "You know what your problem is, Professor Ruiz? You are a *tree hugger*. You want to solve the problems of the world with sunshine and fresh air without ever looking at the practicalities. Renewable energy systems are *just not practical yet*." Once again he fell back on the mantra that "the fossil fuel sector is cleaning up its act".

Professor Ruiz looked shocked. Daniel realized that he had crossed a line. He wanted to apologize for the personal attack, but the host was now winding up proceedings.

"And so, dear viewers, I'd say we haven't found a clear winner tonight. I have my own personal beliefs about this topic, but I will let you decide based on what you heard. Thank you for watching, and good night." The Professor offered a terse goodbye to both the host and Daniel, then walked off pointedly, leaving the two chatting.

---

6. While batteries are the most common type of energy storage mechanism, there are others. Pumped storage uses elevated lakes to pump water up when there is electricity being generated, and release it when needed to drive turbines. Similarly, compressed air in old salt caverns is being used in some parts of the U.S., while in some wind systems these are being paired with quick-start natural gas turbines. Then, of course, there is the electric grid.

10

José sat across the mahogany boardroom table from Daniel and Miguel. Miguel spoke first.

"Daniel, that wasn't a bad performance."

José smiled, and stirred his coffee. "I agree. I think you got a few home truths across about renewables," José added. "I especially liked your savaging of Ruiz. Serves him right for being a thorn in our side."

But Daniel was beginning to have second thoughts. He hadn't felt proud of himself.

"We need you to strike while the iron is hot. Let's work on discrediting this Professor Ruiz once and for all. I want you to go on campus talks, and write more guest editorials," said José. "Our shareholders need to see an increase in market share. I don't see Ruiz helping in that regard."

Daniel nodded in agreement, but something was making him feel uncomfortable.

"OK, let's see some renewed attacks on Ruiz this week. He hasn't really let up on his poisonous opinion pieces, so we have to go on the offensive," said José. "I've got a meeting to go to now, so make it happen, you two."

11

The next day, Maria was with the Professor in his office. "Daniel was pretty much all over you in the end," she said. "But some of it crossed the line."

"Yes, I know. But he's just singing Acme Oil's tune. I know about José Garcia. He won't be happy until my reputation's shot," said Professor Ruiz. "I know quite a few oil industry execs, and though we have our differences of opinion, it's still civilized. But not José Garcia. He takes every attack on the oil industry personally."

The Professor pulled himself up and concentrated on matters in hand. "How are we on the guest editorial on wind energy for the *National Journal*?"

Maria was working on it; she just needed to pull some more data together. Ruiz took a copy of an academic journal from his bookshelf, and flipped through some of its pages, clearly looking for something. He finally handed the journal to Maria, pointing at a page.

"Here, use this. It's an example of how some European countries, by mandating a reduction in emissions, have actually been able to recoup their costs and develop clean technology industries and new jobs, in the face of widespread belief that spending on renewables would slow down growth," said the Professor.

"Thanks," said Maria, grabbing it and bundling things into her bag. There was a lot to do before Tracy's birthday party.

12

Pedro Dominguez hadn't changed from the days when he and Daniel used to sit in high school together. In fact he looked faintly incongruous in his expensive suit. But Pedro was pulling in a serious wage from Eagle Oil.

"Daniel, I've been watching you," said Pedro over a coffee after work. "It seems like your whole *modus operandi* these days is *ad hominem* attacks on your old professor."

"Er, when did you start speaking in Latin?" quipped Daniel.

Pedro grinned. "But the thing is: he does actually have a point on some issues. And *we* have a point on some issues. But it doesn't mean we can't work on it together. In fact, did you know that we have a subsidiary – Eagle Renewables? It's one of our fastest-rising subsidiaries. You see, at Eagle Oil we've come to view the solution to the total energy supply problem as a slow but deliberate move from fossil fuels to renewables and other clean energy sources.

"Renewables *are* slowly going to replace some of our energy needs – maybe not soon, but expect it to gain more share into the future. Daniel, your friend José is one of the last bastions in the industry still fighting for a purely fossil fuel future.

"Plus there are transition technologies. They improve on the efficiencies of fossil fuel technologies, like the Brayton-Rankine combined cycle powerplants which use both a gas turbine and a steam turbine together so that less heat is wasted. Unlike older simple-cycle powerplants whose output efficiency is limited, these combined-cycle power plants can reach efficiencies of up to 60%, generating more electricity and reducing the wasted heat from the same amount of fossil fuel as compared to older types of plants."

"Yeah, maybe," said Daniel. He was still in thrall to his new employer José's way of thinking, but there was something about the way his boss aligned himself against Professor Ruiz that troubled him.

"You know what you need, Daniel? You need to lighten up and spend some time with people your own age. It's Tracy's birthday party tonight. You remember Tracy? Starts with food at seven. Should be a good one. Why don't you come?"

13

Maria arrived late. There were just a few people, mostly students and some faculty members. No Professor Ruiz, but she recognized some of his students.

"Maria! I'm so glad you could make it," said Tracy, kissing her on the cheek. Pedro was beside her, his hand tucked inside her upper arm possessively. Tracy put Maria's gift on a pile with the others on a table in the hallway, but then she and Pedro grabbed Maria's arms and led her into the dining room.

"Maria," said Tracy excitedly. "Can we set you up on a date with someone?" She ignored Maria's groan of protest. "He's actually at the party tonight. We'll introduce you. He's a nice guy!" Pedro was nodding and smiling encouragingly.

"Honestly, you really don't need to do this. I'm perfectly fine as I am – thank you," said Maria defiantly.

But Tracy and Pedro weren't backing down. As they led her into the kitchen, Maria saw a young man with his back turned to them, but who looked familiar.

"Maria, we'd like you to meet..." began Tracy. But at this point Daniel turned around and Maria gasped. "Daniel Vargas! I *don't think so*. Not in a million years," said Maria with real passion.

"The feeling is entirely mutual," countered Daniel quickly, trying to retain his dignity. "I haven't really got time for tree-hugging hippies."

Tracy wasn't giving up on the date idea and leapt to Maria's defense. "Daniel, Maria really is an amazing girl, and I *really* think you should give this a chance."

Pedro chimed in. "And for your information, Maria, I've known Daniel for years. Daniel is a seriously great guy. You should give him a chance." The persuasion and cajoling continued for a good five minutes until it seemed so ludicrous that all four of them were laughing. "OK, you win," said Daniel, and with mock courtesy: "Maria. I would be honored if you would join me for dinner next Thursday." To which he added mischievously, "I suppose it would have to be somewhere vegan?"

14

Professor Ruiz actually laughed out loud when Maria told him. He took a slurp from his coffee mug. "Daniel Vargas. My my. Now didn't I predict this? You might be the making of him. See if you can rescue him from the dark side."

Maria sat down next to him and grabbed his newspaper rudely. "I doubt it. I'm only going out with him as a favor for Tracy and Pedro," she said, trying to sound put-upon.

"That's how it starts … Well. Anyway. Great guest editorial in today's paper. You got the food versus fuel concept across very well," said the Professor.

She emptied the contents of the coffeemaker into her cup and shut the machine off. "I don't think enough people appreciate the facts. Yes, if a country can meet the food needs of its people, maybe any extra crops could be converted into fuel. But it rarely works like that and just drives up the price of food instead," said Maria.

Maria had argued that a better alternative would be to produce cellulosic biofuels. These were the biofuels derived normally from the non-edible portions of food, like cornhusks and stalks. It would need some more research and development, but at least cellulosic biofuels would not threaten the food supply. Many scientists were looking at this area, and results were starting to become promising. There was even some research in using algae as a source of biofuels, and even biofuels from wood through termites was being looked at.

"So where is he taking you?" Professor Ruiz asked.

"It's more a case of where I plan to take him," answered Maria.

"Okay, I'll stop being nosy. Just have a good time."

15

The rain was becoming heavy as Maria stood tensed against the cold under her umbrella in front of her dormitory, waiting for Daniel to arrive. But he wasn't late. After a few pleasantries, she was sitting inside a flash-looking red Maserati. Oh dear: this boy really has taken the corporate dollar, she thought. Daniel suggested they see a movie and grab dinner afterwards.

But Maria had other plans. "I don't know. I was wondering if you fancied driving across town to East Bankside," said Maria.

"Why do you want to go *there*? That's a horrible part of town," said Daniel.

But Maria insisted, and Daniel, puzzled, grudgingly complied. He'd had misgivings about this date to begin with and he was already beginning to regret it. They drove past the city center, past the Acme Oil offices, then past the government complex and the city park. Finally they crossed the imposing steel-girder bridge over the polluted river which divided the city. Garbage was piled on the streets; graffiti covered the walls.

Maria spoke only to give directions. Finally, she asked him to pull up beside an apartment complex. It was a gated community, isolated from the city blocks.

The rain had stopped but the streets were still glistening wet as Maria stepped out of the car. She beckoned Daniel to join her before ringing the doorbell. The gate opened, and an older lady with graying hair greeted Maria.

"Auntie, this is Daniel. Daniel, my Aunt Karen," said Maria. Daniel was nonplused by the turn of events but was too well brought up to do anything but be polite as a guest in a stranger's home. Aunt Karen disappeared into the kitchen to get some snacks.

Suddenly, a young boy ran out of one of the bedrooms and dived on Maria to give her an enthusiastic hug. "Hey Juanito," she laughed. "Juanito, this is my *friend* Daniel," she said putting a strange emphasis on *friend*. "Daniel, this is my nephew Juanito. How old are you, Juanito?" "Seven," mumbled the boy shyly. This was turning into a very strange date indeed. Looking for something to talk about, Daniel said, "Good kid. Erm … is he … OK? He looks a bit pale and sick."

"Juanito is asthmatic. All the doctors he's been to say it's because of the air pollution in this area," said Aunt Karen as she returned with a tray. "He's been off school again today with another attack."

"It's because of that old coal powerplant further up the river," said Maria. "It dates back to a time before most of the newer cleaner coal technologies took effect. It's been battling against regulation and refuses to upgrade its facilities. Claims it costs too much. So now many of the kids here are ill with chronic problems like asthma and other autoimmune diseases."

Aunt Karen added: "I suppose that's how business works. They save money. And they've got the political clout to keep it that way. But what about the real costs? What about all the children in the area with asthma and other diseases?

"When my husband was alive, God rest him, we fought them tooth and nail for years. But I just haven't got any fight left in me now. If this was America or Europe would they be allowed to get away with it? I don't know. But they've gotten away with it here. All I want to do now is to get out. Go to another town. If I could afford it …" She tailed off, looking anxiously at the door of Juanito's bedroom.

"Now, Maria, you and Daniel better get moving. I thought you were supposed to be on a date? Why don't you go see a movie or something?" They said their goodbyes at the door of the apartment, and made their way out of the complex.

"Sorry," said Maria in the car. "Was that a bit obvious?"

Daniel smiled. "Well, yes, it *was* a total set-up. Look, for what it's worth, I'll admit that fossil fuels have contributed a lot to environmental pollution, but the coal and oil companies are *cleaning up their act*," he said, reverting to his tried-and-tested mantra. "Newer coal plants are implementing cleaner coal technology, and a lot of the fuels on the market now are the cleaner-burning ones."

"Yes. And that's a good thing. Of course. But isn't it about time that society aggressively – and I do mean aggressively – finally makes a push for renewable energy?" said Maria. "By 'aggressive' I mean significant percentages of total energy used should be coming from renewables. Massive adoption requires significant government and industry partnerships and investment."

Daniel was silent as he drove past the graffiti-strewn walls and piles of garbage bags on the streets.

Just before the railroad crossing that preceded the steel-girder bridge, they both couldn't help but notice the empty oil barrels dumped near the river bank. The barrels were marked ACME OIL.

"Look at that! What a positive effect your precious Acme Oil is having on our environment," scoffed Maria. "So you just dump your empty barrels where you feel like it? No wonder the river is so polluted." As Daniel started towards the bridge he saw the rainbow shimmer on the river, which could only be slicks of oil from the overturned barrels.

"Stop the car!" shouted Maria. Daniel was startled by her outburst and complied, pulling into a sandy embankment. Maria rushed from the car before it had even come to rest. Daniel followed. Maria was staring at the barrels. Most were empty. Some were haphazardly stacked, and some had tipped over, causing the residual oil to seep into the river.

"Look, there's still time to catch the movie at the mall," said Daniel a little desperately.

"Just take me back to my dorm, please," was all she said.

The remainder of the drive was in silence. It was already dark when they reached Maria's dormitory. She was subdued, and calmer.

"Thank you for coming with me to my Aunt's. I'm sorry it turned out this way," she said. "I hope your Acme Oil salary buys you lots of nice things," she said, more sadly than sarcastically.

Daniel could offer no reply and watched her close the door behind her. He shook his head and turned back to his car.

16

José was pacing. It was 8.50 am on Monday and Miguel and Daniel were due for their usual 9.00 meeting with him. They weren't looking forward to it. He must surely have read Maria's piece about fuel crops. And he wouldn't have liked it.

But it was far worse than that. He threw the Sunday edition of the *National Journal* on the table. On page 2 was a news story, the headline of which shouted "ACME OIL TAKES SHINE OFF GOLDEN RIVER: Oil giant faces pollution charges".

"Daniel! What have you been doing these past few days? You were supposed to be giving me happy stories. This is *not* making me happy. Right. We need a response. And fast.

"Miguel, I want you to send a team there this evening to clean up the mess. We need to find somewhere else to dump the empty barrels. Clean it up tonight. Then tomorrow, Daniel, I want *you* to issue a press release denying the allegations," barked José.

"You want me to lie?" asked Daniel incredulously.

José looked at him. "How will it be a lie? Have you seen the barrels? No. Go there tomorrow: will you see any barrels? No."

Daniel said nothing. He wanted to get out of that office as soon as he could. How was he going to do his job without lying to the public? But he certainly didn't feel like telling José where he'd been last night.

"Lies, all lies! And I know where the story's come from: Ruiz! Miguel, I want you to call that TV station again. Tell them that we want another live debate. Once and for all let's demolish this nutty Professor *and* his 'clean energy' ideas."

"And you," José said to Daniel, "you better get some more op-eds published, and prepare to go in front of the cameras again."

It was going to be a long week. Daniel was starting to dread these Monday mornings.

17

The guest editorial in the *National Journal* that Monday was entitled PEDDLING LIES. The byline was Daniel Vargas.

"I've just read another personal attack from your boyfriend," said the Professor calmly sipping his coffee. "He says that the river pollution story was a lie cooked up by Acme Oil's opponents." After a moment's pause he added, "It must have been a really bad date."

Maria was furious. "He's lying and he knows it. He was with me. We both saw the barrels."

"Well there are no barrels now, Maria," said Professor Ruiz. "They must have sent a clean-up crew. You know, Acme Oil are really in a class all by themselves. I can't think of any companies – even the other oil companies – who behave like this."

At that moment a motorbike courier wandered in and left an envelope on the front desk. A colleague called out Professor Ruiz's name. All that was in it was an official letter. "It's another invitation to go on television. Me versus Daniel again – next week," said the Professor.

But Maria had other ideas.

18

Maria was browsing in the bookstore. Like a lot of her friends she treated it like a library. She was just sitting down with a cup of coffee and a paperback when someone slid into the seat opposite.

"That last date was a big success. Fancy another?"

"You jerk, Daniel! You lied and you know it. And you maligned Professor Ruiz's good name. You were with me. You *saw* those barrels."

"Look, I'm sorry," said Daniel. "I've got a job to do."

"Only following orders, eh? We've all heard that one before. How could you even think about doing something like that?" said Maria.

"I have been thinking about it. A lot," said Daniel. "I'm actually having some serious reservations about this job. I don't think I can do it anymore. But first of all, I really think I owe the Professor an apology."

Maria peered into Daniel's eyes. Was he being sincere? She had a feeling he was.

"Do you *really* want to apologize to him? As in 'really'?" she asked. He nodded silently.

"Then come with me," she said. She led him the few hundred yards from the campus to Ruiz's office. Since his television appearance, Daniel had become a minor celebrity on campus. And here he was heading towards Professor Ruiz's office? Something must be going on. Before they arrived they'd drawn quite a crowd from among the students.

Professor Ruiz was in his room, behind his desk. He looked up to see Maria marching towards him with Daniel close behind. Everyone had stopped what they were doing and were staring.

"Daniel, to what do I owe the pleasure of this visit?" Professor Ruiz said with a hint of sarcasm in his voice.

"I wanted … That is I thought … I'm here to apologize to you." Daniel stammered, then added a hesitant "Sir." "I came here to say sorry. I knew about the Acme oil barrels. And I shouldn't have said what I said about you."

"Are you serious?" questioned Professor Ruiz.

Daniel looked at him, then looked at Maria. "Yes, sir, I'm serious."

"But what about Acme Oil? Are you just going straight back to the office to do more of the same?"

"I'll tell you what I'll say to Acme Oil. I'll say they can accept my resignation," Daniel replied.

Professor Ruiz looked at Daniel again, this time more sympathetically. Daniel looked defeated. "And what will you do then?" asked Professor Ruiz.

"I actually don't know, sir," said Daniel. Professor Ruiz looked at Maria. Maria smiled at the academic.

"Daniel, we could really do with some help round here if you're not busy," said the Professor. Maria was still smiling.

"Are you sure? What could I do? With my track record? What would people think?"

Maria held his arm. "I think you will be just what we need," she said.

Daniel nodded. "Yes. OK. Let me think about it. But first I've got unfinished business at Acme."

The three walked out the door. Outside their office, a large crowd of students and faculty had gathered. "Friends!" shouted Professor Ruiz, "We have a new recruit!" There was a spontaneous burst of applause. Daniel was still feeling too confused and emotional to say anything. He just walked towards his car slowly, with Maria by his side. A couple of male students patted him on the back as he left.

But one bystander was not amused by the turn of events. He had some unwelcome news for his boss at tomorrow's 9 am meeting.

19

When Daniel got to the Acme Oil office he was told immediately that José wanted to see him. In the boardroom the CEO was sitting alone at the head of the table.

"Sir, I have something important I want to say to you…," began Daniel.

But José cut him off abruptly. "SIT DOWN! Don't forget, Daniel, that Miguel gets around, sees a lot of things."

Daniel sat down and waited as José took off his suit jacket and rolled up his shirt sleeves. He walked towards the window and stared out across the city.

Nothing more was said so Daniel spoke. "Sir, I can't do this anymore. I'd like to tender my resignation effective immediately," said Daniel in a soft, quivering voice.

José looked at him with a patient, paternal expression. "Now look, Daniel. I know what you're going through. I was an idealistic young man just like you when I started this company." Daniel was surprised to learn that José Garcia had started out simply as an oil trader, and worked up the ranks until he had his own money to start his own oil-trading firm. From there he branched out into distribution and retail, until he built his own multi-million-dollar empire.

"Don't get me wrong, Daniel. I think renewables are all well and good, but they are simply too expensive and unreliable to pursue. I'll grant you that there are still a few accidents here and there, but oil is and will still be the main basis for running our economy. Think about the improvements in refining and distribution, and the emissions control regulations that the industry has signed up to."

"But what about those barrels in the river? Was that ethical? Responsible? Oil companies don't have to operate like that. And plenty of them don't."

"Now come on, Daniel. You and I both know that if we were to observe every regulation on the books, dot all the *i*s, cross all the *t*s, we simply won't make any money. So some regs you observe, and some you don't."

Daniel was so taken aback by this that he didn't speak. Breaking the silence, Miguel burst in. "So here's the traitor! I can't believe you're still here," he snarled.

José motioned him to sit down. Daniel stood up. "I'm really sorry. I can't work for Acme Oil anymore. Again, sir, thank you for offering me this job. But I have to tender my resignation effective immediately," he said.

"Get the hell out of here! Go do whatever you want!" snapped José.

Daniel walked out, his heart beating loudly and both men glaring at him. "So what now?" queried Miguel. "Who's going to do the head-to-head with Ruiz next week?"

"I will," José said. "I will."

20

Professor Ruiz was in the middle of giving one of his classes on climate change when Daniel walked in. After a small disruption Ruiz continued with his lecture.

"And so while we should admit that there *are* forces in nature that contribute to climate change, in reality a large part of the cause is anthropogenic. Powerplants that burn fossil fuels emit greenhouse gases that pollute the atmosphere and cause climate change. Even cleaner coal plants still have to deal with the issue of how to contain carbon dioxide.

"What cleaner coal technologists are trying to do, for example, is contain the carbon dioxide emissions in underground caverns. But nobody has yet tested the long-term success of this. Others are trying coal gasification, to burn the cleaner resulting gas instead.

"Even agriculture contributes to climate change. The manure from cows, for example, releases methane into the atmosphere. This is why farms that use their biomass and animal manure to generate electricity can get carbon credits, because agricultural manure has been identified as a significant contributor to climate change."

Daniel thought about Juanito. He thought about the state of the city on the other side of the bridge. And he thought about the impact of fossil fuels on climate change. He looked out of the window and saw a dark smog cloud shrouding the city in the distance.

Ruiz is right, Daniel thought, about the hidden costs of using fossil fuels. I've been a fool. It doesn't matter that fossil fuels are getting cleaner. It's too little too late. Society has to get seriously – not just in a token way – into clean energy. It is the only way that costs will drop, if governments adopt mandates to encourage research and early investors.

"Now using fossil fuels, although it may seem like the cheapest alternative right now, is actually expensive in the long run. We are poisoning our environment, contributing to climate change, contributing to increased healthcare costs due to diseases like asthma, chronic obstructive pulmonary disease and bronchitis," continued Professor Ruiz. "If you add that all up, you'll find the costs are a lot higher than commonly thought.

"Plus, there are the economic implications. Remember: a lot of these fossil fuels are imported. So a country's earnings have to go into purchasing fossil fuels, which are an expense, instead of going into investments. While renewable energy might be more expensive now, a lot of that is in upfront capital costs. You can view these upfront costs as investments. The 'fuels' you need to run solar, wind and biomass systems are basically free or things we throw away. Sunlight is free, wind is free, and biomass is just being thrown away."

At this the bell rang and the students began to pack up and leave. "Professor, I've done it. I've resigned. Is the offer still on?" said Daniel.

"I think you've made a very courageous decision there, Daniel," said the professor. "And thank you for coming to my class. I guess that makes you unemployed right now. In which case, I'd like to introduce you to someone. Come and see me at 8 am tomorrow."

"OK, I'll be there," said Daniel.

"By the way. Maria is in my office right now, if you want to see her," said Professor Ruiz.

Daniel just smiled.

21

"OK, we're going on a little drive to see my friend Hans," said the Professor the following morning. "He runs a clean energy company. And he needs a guy with some public affairs expertise."

Waiting for them at the door of the Clean Energy Inc. building was Hans, a bespectacled man in casual clothes, smiling warmly. Hans had been a power industry executive some years ago. His house, which also served as his office, was an ultra-modern all-white marble building, with open spaces and Zen gardens.

Daniel saw the wind turbine on a nearby slope, spinning busily. He wondered how much power it was generating. And he saw the roof, bejeweled with solar panels.

"Hello Hans. I'd like you to meet Daniel," said Professor Ruiz.

"I've heard a lot of things about you, young man," said Hans, shaking his hand vigorously. "Anyway, come inside and let's talk."

Hans paused at the foyer. "I'd like to show you something," he said. He opened a small door that revealed what looked like a panel board with a digital meter. "See that? Right now, my house is generating more electricity than I currently need, so I am selling power back to the utility company."

He motioned them to proceed to the dining area. In the corridor, Hans pointed to his left. "Over here, if you look at this air conditioner, you'll see it's not an ordinary air conditioner," he said. "It has a microchip which allows me to specify at what temperature it turns on and when it turns off. I can also program it to only turn on when there is someone in the room, as there are also room sensors here that detect human presence."

"Many of the appliances in this house are what we call 'smart appliances', which means I can program when they go on power-saving mode or turn off. Besides which: even the power company can set them remotely when needed."

"But how are consumers supposed to afford new appliances like these?" asked Daniel.

"Slowly but surely, appliances will start to incorporate these smart features. In terms of financing, the power company might even get in on the act and help their customers," said Hans.

Daniel was confused, and Hans could see it in his face. "You see, with the current crop of appliances we have, the electric companies have no way of knowing when we will use it or how we use it," said Hans. "The power companies need to build their power-plant capacity to exceed what they estimate would be the peak demand, which usually happens only during certain hours of the day. Just to meet this demand, power companies have to keep an extra set of generators spinning. This is what the power industry calls the spinning reserve, and a lot of those still run on fossil fuels."

"So with smart appliances like these, the power company can reduce the number of spinning reserves that it needs…"

"Because it can predict the demand!" Daniel finished the sentence.

Hans nodded. "Yes, you've got it. Plus, instead of building more carbon-spewing powerplants, the power company can instead tap citizens or companies who want to sell power back to it, to be its spinning reserve. That is the trend in places like California these days.[7]

"Initially, much can be achieved with energy efficiency efforts with cars, houses, buildings and other energy-consuming devices. These can waste a *lot* of energy. Although each device has what is known as a theoretical maximum efficiency, often there's a lot of room for improvement in energy efficiency."

Daniel was impressed. He had never seen a house like this before. Even the lights used in the house were LED-based, and very energy-efficient.

Hans led them into the dining room, where food was already waiting for them. He motioned to them to take their seats. The dining room had a pleasant view of the surrounding fields, including the wind turbine that was still turning briskly.

"So, Daniel, I heard you've parted company from our friend José Garcia over at Acme Oil. That is a smart move. José is one of those who made their fortunes in the old era when fossil fuels were king. He cannot appreciate that clean energy systems like the ones I install will be capturing some of his market from him," said Hans.

The conversation continued over lunch. "You see, all these things are expensive now, but they will eventually come down in price once the market decides that they want them. But without early adopters like me, and some of my customers, the price will never drop. We will always be waiting for that time," said Hans.

Professor Ruiz tapped Daniel on the shoulder. "This is why it is important for governments and societies to embrace clean energy. With a combination of mandates, subsidies, support for research

---

7. In August 2004, California Governor Arnold Schwarzenegger announced the million solar roof initiative, which seeks to lessen their dependence on fossil fuel sources. See http://gov.ca.gov/index.php?/press-release/3588.

and development, and early adoptive customers, the price will eventually fall."

"Daniel," said Hans, "I've seen you on television. You're a natural. You're young, articulate and persuasive. Let me cut to the chase. I'd like you to work for me."

"Yes, sir. Count me in." They shook hands, with a beaming Professor Ruiz looking on.

"So, Professor, I guess all you have to worry about right now is your televised debate with José Garcia next week," said Hans. Professor Ruiz looked at Daniel and grinned.

22

José and Miguel were in a gloomy corner of a coffee shop near the university, planning their strategy for the debate. Miguel saw Maria entering. Miguel made it his business to know what was going on and who was connected to whom. So he knew exactly who Maria was and what she was up to. With José's back turned to her, she was oblivious as to whom she was sharing the café with. Maria chose a seat near the window with a view of the street corner and the entrance to the campus, and took out a book.

"I have an idea," said Miguel and pushed back his chair.

"Hi. Aren't you Maria? I think I recognize you. Didn't I see you at one of the campus talks on renewable energy? What's his name? Professor Ruiz I think?"

Maria looked up from her book and smiled encouragingly at the slight middle-aged man in a suit. Miguel continued: "I really like what I'm hearing about clean energy in academic circles. It's high time the fossil fuel industry was challenged."

Maria said nothing but smiled and let him continue.

"I've heard about that TV debate between Ruiz and Acme Oil. I really hope the Professor demolishes them."

"You seem to know a lot about it," said Maria.

"Well, I happen to know Daniel Vargas. He's a … friend of the family. And I've heard that he quit his job at Acme Oil in order to grandstand against them – which I think is excellent!" he added quickly.

Off her guard, Maria said in a confidential whisper, "Acme Oil never comes to these discussions with enough data, and we're always well armed. If we push them for data we'll slay them."

"Well, that's good to know. I really hope you do well next week. Pleasure to meet you. Good luck!"

"Boss, let's go before someone sees us. We've got what we need," he said to José as they drained their cups and gathered their papers. Maria was too engrossed in her book to notice.

23

On Daniel's first day at Clean Energy, Inc. Hans invited the young man for a walk around the property to see some of the electrical connections. The wind was not blowing that day, and the skies were cloudy. A good day to make a point about the realities of wind and solar systems.

"So there will be some days when there's no wind or sunshine available, or at least nothing significant," said Hans. "Part of the answer is to utilize whatever options work best for a particular location. There is no one-size-fits-all answer for all locations. Each location may have a different set of solutions."

The wind turbine they were looking at had variable-geometry blades, meaning that the angle of the blades could be adjusted, depending on the wind speed. Connecting the solar and wind systems to the electric grid, through devices called inverters, was part of the answer.

"You can also develop hybrid systems, such as wind turbines and natural gas turbines, to back each other up. As a transition step, compressed natural gas is a relatively clean fossil fuel that produces fewer emissions than regular fossil fuels."

Hans went on to say that some countries like France and Japan even use nuclear energy as the backbone of their power requirements, although politically it would be very difficult to argue nuclear power in some countries, especially with incidents like those that happened in Japan and Chernobyl.

Hans continued: "However, nuclear reactor designs have improved over time. The current ones have passive safety features, such as control rods that drop automatically instead of relying on manual intervention for these to be pushed in. There are also designs such as pebble bed reactors that keep the radioactive material inside a core of hardened graphite. But another concern is how the spent fuel will be disposed of properly."

"So what is my key learning here?" asked Daniel.

"Your *key learning*, Daniel, is that there is no one-size-fits-all solution," said Hans. "You need to be aware of solutions using solar, wind, biomass and other energy sources. A particular technology may work best for one town, but not for another town in a different part of the country."

As Hans stared out at the surrounding countryside, his mind turned to the subject of agriculture and livestock farms. "If a farm has a lot of biomass which is generating methane, for example, people should tap it for energy. Rather than simply pushing clean technology blindly, you should advise your clients to live more efficiently and exploit whatever is abundant in their area for their energy needs. Some types of algae can even be tapped as a source of oil. Even waste from landfills can be converted into energy. Processes such as pyrolysis can convert waste into solid char and syngas that can be used to drive electric generating systems. Some countries even use the salinity gradient between salt and fresh water to generate electricity.

"So, Daniel, are you getting the hang of this?" asked Hans.

"I think so."

"Daniel, I think you might be ready to meet with a potential client from out of town."

I think I'm going to enjoy this job, Daniel thought to himself.

24

In the hybrid car that Hans had lent him, Daniel left early the next morning to meet his first client. He stepped on the pedal lightly, accelerating the car to overtake a cargo truck on the highway. "Who would've thought I'd be driving around in this?" he mused.

Hans had explained to him that the gasoline engine would only kick in at the start and then occasionally to recharge the battery. But most of the time it was the electric motor doing the work.

Daniel zipped along at 50 mph on the freeway. He could see the big coal plant near Karen and Juanito's house. The car was handling very well, even at that speed.

They should close that plant down, or at least upgrade it with cleaner coal technology, he thought to himself. He thought about Juanito's pallid face and hacking cough.

Just then he noticed some police lights flashing and eased off the accelerator. It must be an accident, he thought. Stepping on the brake gently, he could hear Hans's explanation replaying in his head: it was the motor itself that did part of the braking. By using a process that engineers called regenerative braking, the motor was in fact acting as a generator. Whenever he stepped on the brake pedal, the motor actually took that as an opportunity to recharge the battery. Wow, he thought as he put it into practice, that is really neat.

Hans had said that, eventually, battery technology would improve to the point where you no longer needed the engine. The car could be charged and operated purely as a battery electric vehicle. Some people even proposed that during certain hours, when some of these electric vehicles were not being used, they could be used as storage for peak load equalization and intermittent source storage.

The accident on the highway was minor and it appeared to have been cleared up. The policeman waved him through and the traffic sped up again.

His client lived on a farm on a dirt road a few miles from where he exited the freeway. "Paul?" asked Daniel of the man in overalls. "You must be Daniel," replied Paul.

Daniel found himself sitting next to Paul on some bales of straw beside the stables. "I understand that you have a lot of animal waste that can be put to good use. Methane for power generation," said Daniel. "We can also add some solar panels, or even a wind turbine, if it makes sense, to augment your methane gas system. And, if you're already connected to the grid, you can sell your excess power back to the utility. They can just subtract the difference from your electric bill."

Daniel was delivering his pitch with enthusiasm. "Aside from looking at clean energy options, you could also have an energy efficiency audit done. Many homes and buildings are inefficient and waste a lot of energy. If you add up all the electricity wasted like this across the country it would be equivalent to a new power plant."[8]

---

8. Energy efficiency is actually one simple and easily achievable way to combat climate change. The equivalent power plant (EPP) concept is a powerful way to illustrate its value.

"I think I like the sound of the grid connection and the energy efficiency audit options. Keep talking," said Paul. Daniel explained the economics of connecting to the grid, just as Hans had explained it to him.

It was an hour later when Paul and Daniel were shaking hands. "Well, you've sold me on the system! I think I can live with that last price you quoted. We can start whenever you like," said Paul. "You know, I never thought I'd see the day when all this animal waste would actually make money. It all makes a lot of sense!" said Paul.

"Yes," said Daniel earnestly, "it really does! Let me get back to you with a schedule so we can get started as quickly as possible."

Daniel was singing along to his car stereo as he drove back into town, and drumming his fingers in time on the steering wheel. He'd clinched his first sale for Clean Energy, Inc. And it felt good. And what's more it felt *right*.

He was jolted from his reverie by a cell phone call from Maria. Juanito had suffered another asthma attack and she was with him in the emergency room.

"I'm on my way," he replied.

25

Maria and Juanito were seated on one of the benches in the ER. Juanito was inhaling through a nebulizer. Daniel noticed that there were a lot of kids in the room that day.

"Why so many kids?" he asked.

"It's the coal plant again. There was a big emissions problem. Lots of black smoke. It's affected a lot of kids; they got sick really quickly. Anyway, the Mayor's asked the plant to shut down temporarily," said Maria.

"I know what's going on here," said Daniel. And I know who's to blame: José Garcia!"

"What? How?"

"I happen to know that he was recently awarded a contract to supply coal to that plant. He's got some investments in the coal-mining sector, but the thing is … Look, to put it simply, the type of coal they're supplying is not a good type of coal."

"I don't get it. You can't get away with it these days, can you?"

"Well, if you mix the two types together and don't tell anyone…"

So you think he is mixing in the substandard coal with the good stuff before it gets to the plant?" Maria asked.

"I don't have any proof, but I'd bet a lot of money on it."

"So how do we prove it?" asked Maria.

"I don't know."

26

"What have you got for me, Miguel?" asked José, pacing back and forth across the room.

"Well, sir, we have all the data. Spreadsheets, pivot tables, pie charts, PowerPoint slides. All you need to underpin an argument *pro* coal and other fossil fuels and *contra* renewable energy," said Miguel. "And here's the 'script' to go with them. We've second-guessed all their arguments and prepared responses."

"OK, leave them with me. You can go." But he sensed that Miguel was hesitant to leave.

"Miguel, I said you can go. What is it?" said José. Miguel looked at his boss uneasily. "Sir, I was just wondering – about the East Bankside power plant. We really have to stop mixing in the substandard coal. Besides, either we don't have scrubbers on that plant, or if we do, we aren't operating them. There are health implications. What about all the k…?"

He didn't get the chance to finish. José banged his fist on his desk. "You keep your mouth shut about that! We are making a *lot* of money on this supply contract. Anyway, why the sudden concern about a few sick kids? Kids get sick every day."

"I happen to know some of the families involved, that's all." What he didn't explain was that his elderly parents lived in that part of town and he was getting increasingly worried about his father's deteriorating health.

José pointed his finger at Miguel. "You just make sure that I win this debate. The powerplant does not concern you. Do you understand?"

"Yes, sir. I understand perfectly."

27

"I can't remember the last time you paid us a visit," chided Miguel's mother.

"I'm sorry, Mom. You know what work's like. How's Dad?" From their window the neighborhood looked the same. Graffiti on the streets. Homeless people begging on the sidewalks. There was a sour smell in the air and a powdery feeling when you breathed it in.

Miguel's mother put coffee and pancakes in front of him. "He's the same, Miguel. He's not been out of bed since his birthday." She looked pensively out of the window. "I wish I never had to look at that coal plant ever again…"

The coal plant was so near to their house it dominated the view from the front window. A little further you could see the billowing smokestacks of the powerplant.

Miguel felt a huge stab of guilt in his chest but couldn't find any words. When Acme Oil had acquired the pulverized-type coal plant a few years ago they had neglected to put in certain contemporary emission reduction features such as scrubbers. Modern coal plants had already reduced sulfur dioxide (SO_2) and nitrogen oxide (NO_x) emissions significantly, but not this one. Miguel felt trapped.

"Tell Dad I was here. I need to get back to the office. Thanks for the breakfast." Miguel kissed his mother on the forehead.

Miguel recalled this neighborhood 20 years ago before the coal plant arrived. It had looked like any other decent suburb in other parts of the city. But the coal plant precipitated a gradual exodus and property prices plummeted.

As he drove back to the office, he felt an unfamiliar sensation. He realized he was crying. He had always told himself that there was nothing he could do about the situation. It was just the way the world worked.

28

Daniel and Pedro were sitting on the tall stools at the bar.

"I've been talking to Maria," said Pedro. "She's been telling me about her nephew. I am *livid* about what's happening out there. Apart from anything else, it gives responsible members of the industry a bad name."

"Pedro, you talk about 'responsible members of the industry'. If you really want to be responsible you should be spending your time working on renewable energy," replied Daniel, sipping his beer. "Not in a token way like you're doing now, but 100%."

"Well look, Daniel, we may have our differences as to how quickly we will get there, but eventually we *will* get there," Pedro replied. "The point is: how do we go green while ensuring a stable supply of energy? There is still a role for clean fossils to transition us in the meantime while we continue to constantly improve both renewable energy and traditional energy."

Daniel leaned over to Pedro and spoke quietly and confidentially. "How do we find out if José Garcia had anything to do with the incident the other day at the coal plant?"

"Well, you could, if you could trace the incident to his shipment of coal to the plant."

"There is something fishy going on in that coal plant, and we need to find out what it is."

"The combination of the poor grade of coal they are using, plus the fact that the plant probably doesn't have scrubbers makes for a deadly mix. Unfortunately, unless someone squeals from the inside, I don't think we will ever know what's really going on," opined Pedro. "You see, after the coal is incinerated, the heat from the incinerated coal converts water into steam in a boiler and drives an electric turbine to generate power. The exhaust should go through scrubbers, and sometimes through electrostatic precipitators, but I'm not sure if this plant is actually doing that."

29

Miguel walked into José's office a few minutes before the meeting. Ever alert, his eyes wandered to the trash next to José's desk. He stooped down to pick out a scrunched-up A4 document with a glossy card cover. It was the latest assessment of cleaner coal technology, or what engineers called Carbon Capture and Sequestration (CCS). There were color brochures in the trash as well, which promoted better scrubber technologies, more efficient burn technologies such as fluidized beds, and combined cycle Brayton-Rankine powerplants. There was another which went into detail about some of the newer technologies like coal liquefaction, which did not burn the coal directly, but converted it into syngas.

A pretty eloquent testimony to José's attitude to climate change mitigation, mused Miguel: creased and thrown into the trash. Coal plants have been identified as one of the leading generators of greenhouse gases, particularly CO_2, responsible for climate change. And CCS has been identified as a potential solution for reducing CO_2 emissions from coal plants. The technologies mentioned in the brochures promised to burn coal more cleanly, or extract higher efficiencies to convert heat into electricity such as the combined cycle plant technology.

Making himself comfortable on José's office sofa, Miguel leafed through the report. He had actually been unaware that roughly one ton of CO_2 is emitted for each megawatt-hour produced by typical coal plants.

José burst in. He didn't greet him, just demanded to know what he was reading. Miguel closed the report and put it on José's desk.

"Boss, isn't it time we looked properly at cleaner coal technology? Either CCS or the other more efficient coal burning alternatives out there like fluidized bed or even coal gasification? I mean, how many times have we said publicly that 'fossil fuels are cleaning up their act'? These are the key cleanup mechanisms that we need to take seriously if we're going to walk the talk."

José took the brochures from the desk and threw them back into the trash.

"Now you listen to me. What I say in public and what I do in private are two different things. Sure, cleaner coal may be the way to go in the future, but the technology and the economics need to advance first before we take it seriously. If we implement carbon capture now, in its current state, it will cost us up to around $30 per ton of CO_2 that we generate.[9]

"We need to be careful about this, Miguel. The carbon tax advocates want us to pay big-time for every ton of CO_2 we emit into the air. That's just not feasible. The technology just isn't there yet, and the cost is too high."

"But what about the future?" asked Miguel. "Can we not even consider coal liquefaction for our plants in future, instead of pulverized coal plants?"

---

9. The $30 figure is the estimate given in the 2007 MIT *The Future of Coal* report at http://web.mit.edu/coal.

"It's too early to talk about CCS," was José's reply. "Believe me, Miguel, I *have* looked into it. If we have to look at cleaner coal, we'd first have to have either fluidized bed plants that burn the coal efficiently in a bed of air and sand; or coal liquefaction, which converts the coal into a liquid fuel that could be burned to run turbines more cleanly just like natural gas. But why do that? The pulverized coal plant is still working just fine. Anyway, we still have to conform to the government's proposed new sulfur dioxide and nitrogen oxide emission levels."

"But boss, if that carbon tax *does* get passed, then wouldn't we be better off adopting cleaner coal technology as soon as possible?" asked Miguel.

José looked at the view of the city from his window. "It won't go through. Our friends in Congress won't let it," he said.[10]

Miguel felt a surge of anger rise within him. For the first time he despised his boss and the kind of thinking he stood for.

---

10. In fairness to legislators worldwide, many of them have been supportive of clean energy legislation.

30

The harsh glare of the studio lights struck the faces of two men staring contemptuously at each other as the studio crew buzzed around them in the last seconds before transmission. It was the same talk-show host who had chaired the first debate. She gave the floor to José first. His opening pitch was that fossil fuels and traditional sources were being given a bad name.

"These so-called dirty fuels are not in fact so dirty anymore. A lot of science and technology has gone into emissions control. And who in the audience – honestly – wants to pay more for their fuel bills? And then find it blacking out every half an hour because the wind's not blowing or the sun's not shining? Renewable energy has been romanticized. The plain fact is that it is not yet mature enough, it is expensive, and it is intermittent.

"There are new developments in cleaner coal technology like liquefaction and CCS, fluidized beds, and better, more advanced scrubbers. These are the technologies that will give society affordable and clean power. Sure, we may have a few solar and wind systems running here and there, but the bulk of our power will still come from fossil fuels. Now and in the future."

It was a strong and well-rehearsed opening. The floor was passed to Professor Ruiz, who seemed less confident, as if he'd been blindsided. "Yes," he started falteringly. "Admittedly, there may still be a role for the clean versions of traditional power sources running alongside renewable sources to supply our power needs for the future…" But he gathered himself and leaned forward with renewed confidence. "But the share of renewable energy will need to become significant. And that will only happen if the government enacts mandates for early adoption and scraps subsidies for fossil fuels. That, together with research and development, will lead to massive adoption of renewable energy systems paving the way for them to become more efficient and cheaper."

The debate carried on like this in a predictable way for several more minutes when the host interjected with her own question. "Mr. Garcia, I understand that your company was recently awarded a contract to supply coal to the East Bankside coal plant, which is now making the headlines for the wrong reasons following reports of dangerous levels of emissions. Can you tell us anything about that incident?"

José was clearly not comfortable with the question but set his face and delivered his pre-rehearsed speech. "Well, first of all, let me apologize to the families, the parents and children, who live in the vicinity. This is an appalling incident and me and my company Acme Oil are at one with you in denouncing it and in insisting that the appropriate measures are taken to prevent anything like this happening again. The coal we supply to the plant is of the highest standard, and its emissions profile is one of the best available. But we're not stopping there. We are currently looking at either fluidized beds or coal liquefaction to clean the plant up." He looked straight into the camera and his voice never faltered. Miguel was off-camera and just stared at the floor.

"And are you willing to submit your coal to tests?" asked the host.

"Ahh, ermm …" José stammered a little while he weighed up his response. "But of course." He threw a meaningful glance toward Miguel. But Miguel was not meeting his gaze.

The host moved on. "Now here is a question that is on a lot of people's minds. If we penalize carbon emissions through a tax or a law of some sort, are we hurting or creating jobs?"

José signaled that he wanted to answer first. "Definitely hurting. It will destroy jobs. The additional costs to business can't help but result in job losses," he said confidently.

"Professor Ruiz?" asked the host.

The Professor responded: "Penalizing carbon emissions might seem to some to be anti-business. But you'd be very wrong. It is actually *good* for business. It will create new industries and new jobs based on an increased need for clean technologies."

It was time for the closing arguments. Daniel, Maria, Juanito, Pedro and some of the other students, professors and alumni from the school were all in the studio audience. Only Miguel was on the set. In an idle moment Maria cast her glance round the studio and started with surprise when she saw him. She recognized him immediately from the coffee shop. And here he was as part of the Acme Oil entourage.

José went first. "Fossil fuels, my friends, are currently an easy target for the liberal media. But the fossil fuel industry is one of the most responsible industries – because it has to be. It has been put under enormous pressure in recent years to clean up its act – and it has responded. There has been massive investment in research and development, in proper disposal, in emission controls. The fossil fuel industry is striving to be a responsible member of society and a guardian of our future and our energy security. We are not, as the slander and simple caricatures often portray us, driven just by profit and growth. We have a commitment to minimize our impact to the environment and to the health of the people we serve."

Professor Ruiz countered: "For just over a couple of lifetimes, we've been locked into a rapacious and ever-increasing race to exploit a finite and diminishing resource that took millions of years to establish itself. In those few short decades we've changed the face of the planet – for the huge benefit of so many aspects of our lives. But at what cost? The pollution and degradation of our environment, and the potentially catastrophic effects of climate change. The evidence is overwhelming. We cannot continue like this. It is our duty to the generations that follow us to quit the fossil fuel habit now and leave what's left of this ancient carbon in the ground where it belongs. Before it's too late. Renewable energy is the only intelligent and responsible way forward.

"Look, I appreciate that there are plenty of well-meaning people in the oil and coal industry who take their responsibilities seriously and are investing heavily in cleaner fuels, and I'm not saying there hasn't been some progress. But while this is happening we all know companies whose record is atrocious, and who plough on maximizing profit without regard for the environmental damage they cause or for public health."

He paused. "So I say, let us look to Parliament. Let's have a mandate requiring by law *by a certain year* that a percentage of our energy needs is sourced from renewables. In doing this, and by investing in research and development, these technologies will see widespread adoption and continuous improvement. And, as night follows day, the price of renewable energy will drop. We will see new industries and new jobs – all based on clean technology."

But there was still a few minutes on the clock, and it was a live broadcast so the host had to keep things rolling. She had a question at the ready. "So, Mr. Garcia, if you are so opposed to mandates and subsidies for the renewables sector, what do you have to say about the large financial subsidies to your own sector? Aren't these just an example of government intervention in the free market? The kind that you so vigorously argue against in the renewables sector?"

José retorted, "Look, without these subsidies, you would all be paying a lot more for your energy."

Professor Ruiz interjected. "I can't believe the hypocrisy of your position! You openly oppose mandates and subsidies to the renewable and clean energy sector while all the time calmly pocketing substantial government subsidies yourselves."

But the exchange was suddenly sidelined as those on set became aware of a commotion in the studio audience. "Juanito!" a woman's voice was heard to scream. The child was breathing desperately and coughing pitifully, his face pink. Somebody shouted for a medic. A male voice was heard to shout "This is all your fault, Garcia!"

José leaned over toward the clip mic on his lapel, "Nobody regrets the health impacts of this unfortunate incident more than I do, but I absolutely insist once again that this is not attributable to the actions of Acme Oil."

Miguel stood up slowly and deliberately and walked across to the centre of the set. The host motioned to keep the cameras rolling; this was television gold dust. Miguel started to address the audience as a quick-thinking crew member pinned a clip mic on him.

"My name is Miguel Contreras. I am José Garcia's chief adviser. I am therefore in a unique position to inform you all that you are being lied to."

José stood up and grabbed Miguel by the arm attempting to lead him away, muttering something no one could hear. But Miguel stood firm.

"This man will poison your communities and compromise your future just for profit. It is he and he alone who is responsible for the coal plant incident. It was his decision and no one else's to mix substandard coal with regular coal, and to deceive the plant, his employees and the community. All to squeeze some extra profit." His voice rose. "He doesn't give a damn about the community or your kids. All those clean fossil fuel technologies he was talking about? He has no intention *whatsoever* of implementing them."

Daniel and Maria were on their feet. The camera was panning round the room in an attempt to miss none of the action. "What about the oil pollution, Garcia?" shouted Daniel. "We were there. We saw it with our own eyes. And then you tried to cover it up and deny all of it." Juanito was now been given oxygen by a paramedic, in plain view of the rolling camera.

José had stopped trying to bundle Miguel offstage; he had slumped back in his chair.

Miguel continued, "This courageous young man is telling the truth. It's not the only incident by a long chalk. José Garcia has been taking shortcuts to bypass emissions and pollution control standards for years.

"If this goes to court, I'm willing to testify," he said looking at José coldly.

"Yes, let's have our day in court," shouted Maria.

"I used to work for him: I've got plenty to say," shouted Daniel.

But there was an empty chair on set. As the room descended into an unintelligible babble of a hundred different conversations, the host spoke to camera to wrap up proceedings.

31

Professor Ruiz folded the newspaper and grinned broadly.
"Well, Maria. What an eventful day that turned out to be."

A week had passed since the studio fracas. The team had been following the ensuing events closely, which had not been difficult as the media had taken a keen interest.

"Professor! Garcia's resigned," squeaked Maria excitedly. "I'm just looking at the news site now. 'Garcia faces criminal charges in connection with emissions violation'."

"When you reach a turning point, you find that events tend to unfold rather quickly," replied the Professor sagely. "I've just received an email from none other than Miguel Contreras, acting CEO of Acme Oil."

"Not another TV head-to-head?"

"Not at all," smiled the Professor. "I've been offered the job as special adviser on renewables strategy. The Board of Acme Oil wants to reposition itself as – and I quote – 'Acme: an energy provider for a sustainable future.' They're announcing publicly that they see cleaner coal as a transition technology and plan to pump millions into renewables R&D. They're already promising to be 100% compliant with emissions equipment and monitoring by the end of the month."

"Wow. What a turn-around. Imagine getting Board support for all that!"

"I think they had no choice but to see the way the wind was blowing. The thing is: it's not just reputation damage limitation – it makes good business sense, as long as you're thinking long-term, which any sensible business should."

"Is this going to put the likes of Clean Energy, Inc. out of business?"

Not any time soon," said a voice next to her. Daniel had snuck into the office without anyone noticing. "We're going from strength to strength. I'm enjoying selling clean energy solutions and I'm doing an absolute bomb in commission right now. It's easy when you really believe in what you're selling."

"Nice one, Daniel. Good for you," said Maria warmly.

She turned and spoke to the Professor. "Have we really hit a turning point?"

"Will all energy soon be clean energy?" asked the academic rhetorically. "It's too early for optimism. We've scored an important victory, but we can't afford to be complacent. The future lies in our own hands. If we stop being vigilant about the sources of energy we use, then we and especially our children, and our children's children, will pay for it. Make no mistake. We cannot simply make the current cost of energy a basis for the argument. If we fail to invest in clean energy, and start adopting it on a proper scale soon, we will end up paying a bigger cost eventually.

"Not until people in general start to make the right choice about energy can we really say that we've turned a corner. Are people convinced yet? Are they willing to make the choice to go with clean energy even if it means paying more in the short term? Or will they continue to ignore the issue and take whatever's offered to them by the fossil fuel corporations?"

"But if enough people make their voices heard…"

About the author

Dennis Posadas is an Asia-based technology columnist, consultant and author whose opinions and articles on Asian technology, innovation and clean energy have been published in *Bloomberg BusinessWeek.com*, *Forbes*, *YaleGlobal*, *Christian Science Monitor*, *South China Morning Post*, *Singapore Straits Times*, *Singapore Business Times*, *Singapore TODAY*, *Japan Today*, *Jakarta Post*, *UCLA AsiaMedia*, *UPI*, *Inquirer.net* and *BusinessWorld*. He has also been an international fellow of the Washington, D.C.-based Climate Institute.

His previous published books are *Rice & Chips: Technopreneurship and Innovation in Asia* (Singapore: Pearson Prentice Hall, 2007) and *Jump Start: A Technopreneurship Fable* (Singapore: Pearson Prentice Hall, 2009). Posadas lives with his family in the Philippines.

Printed and bound by CPI Group (UK) Ltd, Croydon, CR0 4YY

23/10/2024

01777666-0002